Materi

CAN YOU MAKE A
TOASTER
OUT OF
PLASTIC?

by Susan B. Katz

PEBBLE
a capstone imprint

Published by Pebble, an imprint of Capstone
1710 Roe Crest Drive, North Mankato, Minnesota 56003
capstonepub.com

Library of Congress Cataloging-in-Publication Data is available on the
Library of Congress website
ISBN: 9781666350913 (hardcover)
ISBN: 9781666350975 (paperback)
ISBN: 9781666351033 (ebook PDF)

Summary: Plastic is in lots of items you use every day. Find out about
different types of plastic, what they are used for, and if you could really
make a toaster from it.

Editorial Credits
Editor: Christianne Jones; Designer: Elyse White; Media Researcher: Morgan
Walters; Production Specialist: Polly Fisher

Image Credits
Getty Images: LumiNola, 15, PrathanChorruangsak, 11, vitranc, 19;
Shutterstock: BravissimoS, 13, DUSAN ZIDAR, middle right 8, Halfpoint, 14,
i viewfinder, 7, Iammotos, 5, Inga Nielsen, right Cover, Marko Poplasen, 16,
Pixel-Shot, 21, SeDmi, left Cover, small smiles, middle left 8, solarseven, 9,
Tomsickova Tatyana, 17, Yaroslau Mikheyeu, middle 8

Printed and bound in the USA.   4882

Table of Contents

Words in **bold** are in the glossary.

A Plastic Toaster?

Toasters heat food. They use **electricity**. They are in almost every kitchen. They make bread warm and crispy. You can also make waffles and other foods in them.

So would plastic make a good material for a toaster? Let's find out more!

What Is Plastic?

Plastic is a **bendable** material that can be hard or soft. It melts easily. It can be **dyed** different colors. People make plastic from natural materials. Electricity can't go through plastic.

Some plastic is flat, while other plastic is rounded into a shape, like a bottle. Some plastics can be **recycled**.

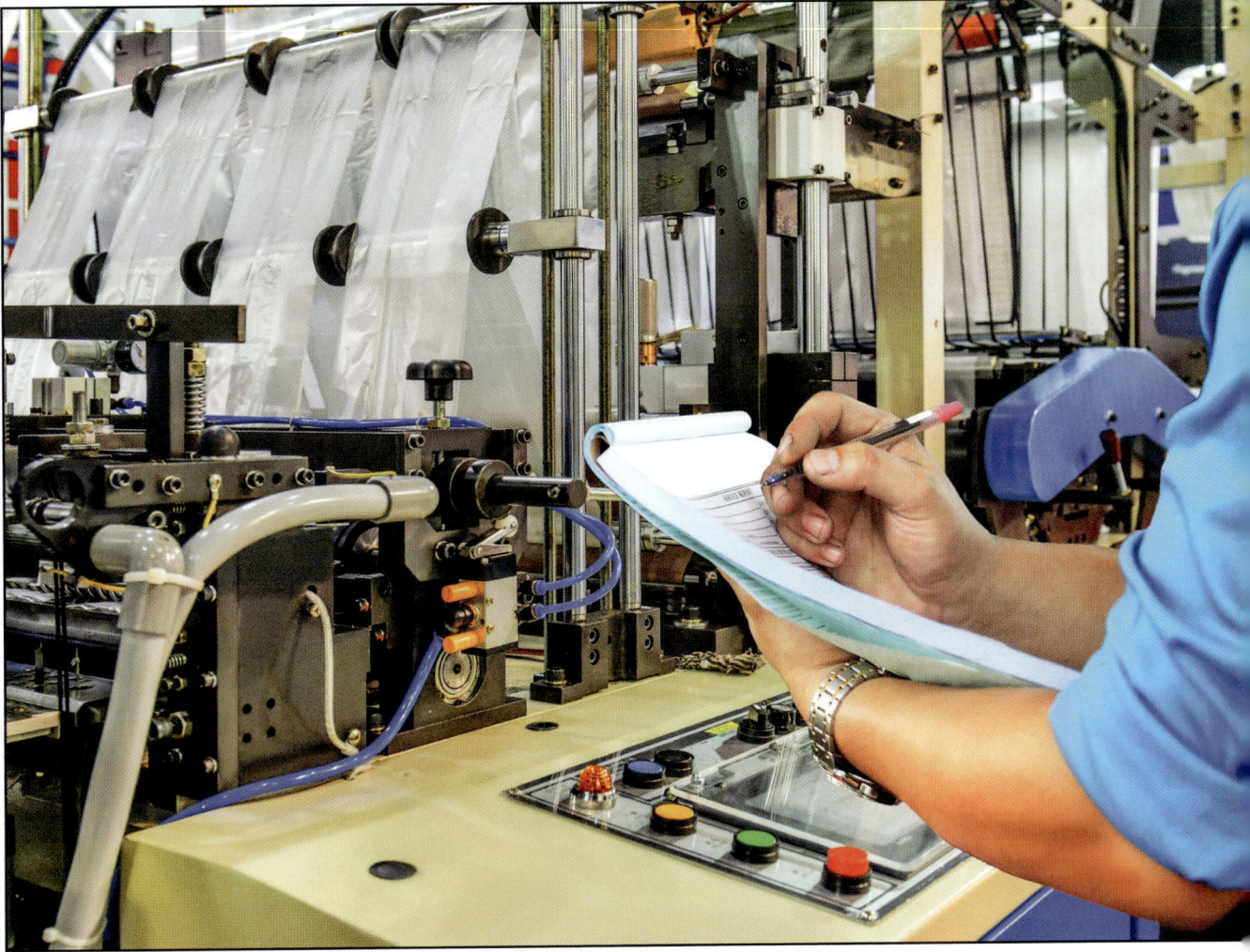

A worker checks the plastic being made in a factory.

Making Plastic

Plastic is made from natural things like coal, salt, natural gas, and oil. They are heated to make different **chemicals**. The chemicals are then mixed together to make different plastics.

coal

natural gas

oil

Plastic is useful. But, sometimes, making it can cause problems. **Toxic** chemicals can escape into the air. Plastic can end up in oceans, rivers, and streams.

Properties of Plastic

Plastics are lightweight. It is easy to make plastic into different shapes and sizes. It can be melted or bent.

Plastic also doesn't break easily or rust. It often can be used for years without wearing out.

Plastic can trap heat. A plastic lid on your hot chocolate works perfectly.

Made from Plastic

Many products that we use every day are made from plastic. It is easy to clean and cheap to buy.

A kitchen has lots of plastic. When you have food leftovers, you often store them in a plastic container. We use plastic bags to store food. Plastic can hold liquids that we drink.

Where else can you find plastic? Laundry baskets, soap containers, and shampoo bottles are plastic. Toothpaste often comes in plastic tubes. Many more bathroom items are made from plastic too.

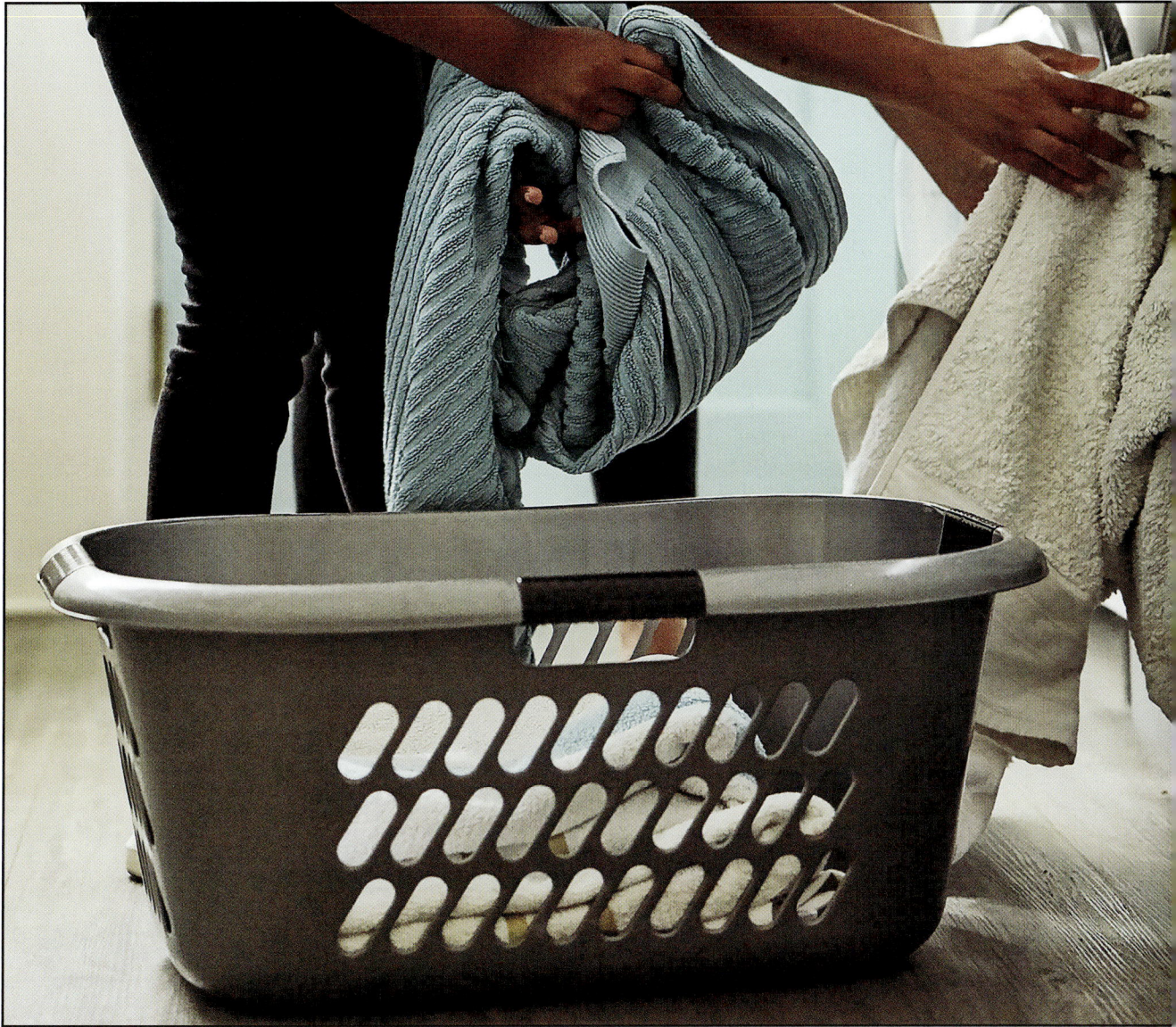

In school, markers, book bins, and trash bins are made from plastic. Some chairs are made from plastic too.

Many of your toys are made from plastic.
From dolls to trucks to your action figures,
plastic is everywhere!

Plastic in a Toaster

You can't use plastic on the inside of a toaster. Electricity would not go through it like it does with metal. Plus it could melt or catch on fire!

But some toasters do have plastic buttons or levers on the outside. You can touch them because plastic doesn't heat up like metal.

So could you make a toaster out of plastic? Would you?

Toaster Time

Draw a picture of the parts of your toaster. Label which parts are metal and which parts are plastic.

What You Need:

- toaster or a picture of a toaster

- sheet of paper

- pencil

- crayons or markers

What You Do:

1. Look closely at the toaster. Decide which parts are plastic or metal.

2. Draw the toaster and color it.

3. Label the parts plastic or metal.

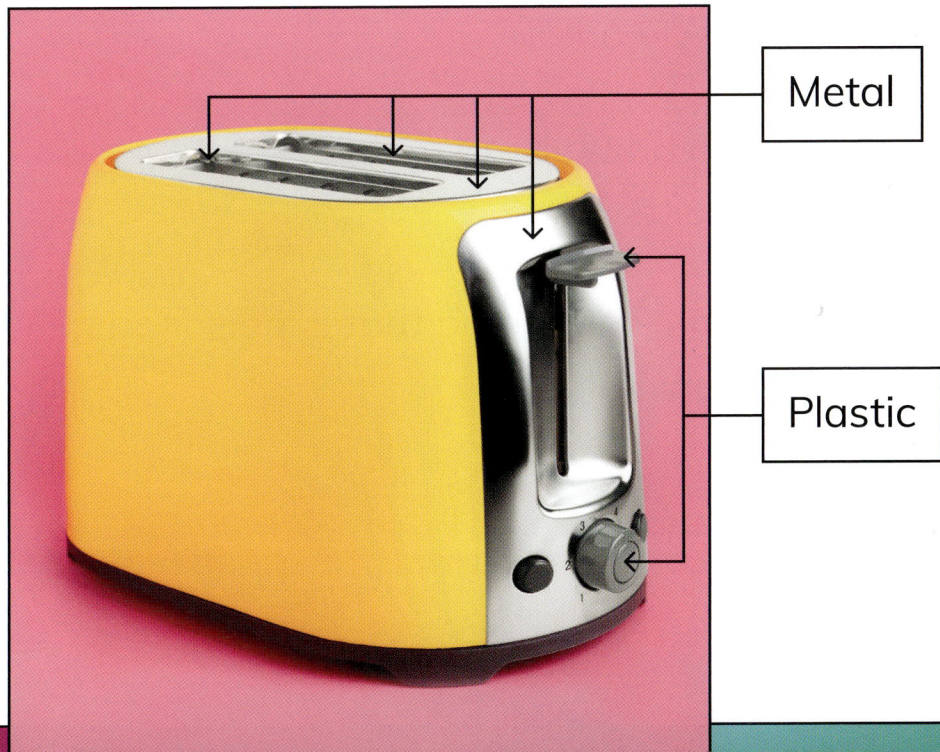

Metal

Plastic

21

Glossary

chemical (KE-muh-kuhl)—a substance used in or produced by chemistry; medicines, gunpowder, and food preservatives are made from chemicals

dye (DY)—to change something's color by adding chemicals

electricity (ih-lek-TRIS-i-tee)—a natural force that can be used to make light and heat or to make machines work

recycle (ree-SYE-kuhl)—to make used items into new products

toxic (TOK-sik)—something that is poisonous for people or animals

Read More

Crull, Kelly. *Washed Ashore: Making Art from Ocean Plastic.* Minneapolis: Millbrook Press, 2022.

French, Jess. *What a Waste: Trash, Recycling, and Protecting Our Planet.* New York: DK Publishing, 2019.

Pagen-Hogan, Elizabeth. *Ocean Plastics Problem: A Max Axiom Scientist Adventure.* North Mankato, MN: Capstone, 2022.

Internet Sites

Britannica Kids: Plastic
kids.britannica.com/kids/article/plastic/400149

DK: Find Out About Plastic
dkfindout.com/us/science/materials/plastics

National Geographic Kids: Kids vs. Plastic
kids.nationalgeographic.com/nature/kids-vs-plastic

Index

About the Author

Susan B. Katz is an award-winning Spanish bilingual author, National Board Certified Teacher, educational consultant, and social media strategist. When she's not writing, Susan enjoys salsa dancing and spending time at the beach.